THE WEATHER REPORT

Natural Disasters

JOHN HOPKINS

PERFECTION LEARNING®

Editorial Director: Susan C. Thies

Editor: Mary L. Bush

Design Director: Randy Messer

Book Design: Michelle J. Glass,
Deborah Lea Bell, Lori Gould

Cover Design: Michael A. Aspengren

A special thanks to the following for their scientific review of the book:

Judy Beck, Ph.D.; Associate Professor, Science Education;
University of South Carolina–Spartanburg

Jeffrey Bush; Field Engineer; Vessco, Inc.

IMAGE CREDITS:

©Paul Seheult: Eye Ubiquitous/CORBIS: p. 4 (front); ©Bettmann/CORBIS: p. 5; ©Bettmann/CORBIS: p. 10; ©AFP/CORBIS: p. 25; ©Annie Griffiths Belt/CORBIS: p. 29 (back); ©Bettmann/CORBIS: p. 31 (top); Associated Press: pp. 12 (left), 16 (top)

Cover: Digital Stock (background), NOAA (front-left), NOAA (front-middle), ArtToday (front-right), NOAA (back cover); Royalty-Free/CORBIS (Stormchaser 2, Vol. 188): pp. 6, 17, 27 (left), 30, 33 (back), 34 (top), 36–37; Digital Stock: pp. 2–3, 4 (back), 7 (back), 9, 11 (top), 12 (right), 13, 14 (top), 22 (back), 35, 40; Photodisc: pp. 26 (back), 38–39; Corel: all art on sidebars, 26 (front), 29 (front); ArtToday (arttoday.com): pp. 11 (bottom), 14 (bottom-right), 23 (right), 24 (right); Hemera Technologies Inc: p. 22 (front); MapArt™: pp. 21, 23 (left), 24 (left), 28; Lori Gould: pp. 14 (bottom-left), 20 (top); NOAA: pp. 8 (left), 15, 16 (middle), 20 (bottom), 27 (right), 32, 33 (front), 34 (bottom); Perfection Learning Corporation: pp. 7 (front), 8 (right); Kay Ewald: pp. 19, 31 (bottom)

For information, contact
Perfection Learning® Corporation
1000 North Second Avenue, P.O. Box 500
Logan, Iowa 51546-0500.
Phone: 1-800-831-4190
Fax: 1-800-543-2745
perfectionlearning.com

1 2 3 4 5 6 BA 09 08 07 06 05 04

ISBN 0-7891-6021-8

Contents

1 Disasters Weather or Not?

When you hear the word *disaster*, what comes to mind? The test you took yesterday? The haircut you got last week? The new shirt you spilled grape soda all over?

While all of these might seem like disasters to you, *real* disasters are much worse. The dictionary defines a disaster as "a sudden event bringing great damage, loss, or destruction."

This can include the loss of life, property, and land. Sometimes disasters leave damage beyond repair. Disasters come in many shapes and sizes, but they all have devastating effects on land and people.

Disasters have many causes. Some are the result of human actions. Others are created by nature. Several natural disasters are caused by weather.

Human Disasters

Humans are responsible for many disasters. Oil spills that flow into oceans destroy animals and their **habitats**. Pollution thrown into rivers, lakes, and oceans damages wildlife. Deadly gas leaking from chemical plants seeps into neighborhoods, making people sick. Raging fires, started purposely or carelessly, burn down forests and communities.

Nuclear power has been responsible for several disasters. In the United States, people living near the Three Mile Island nuclear power plant lost homes when **radiation** leaked into their neighborhoods. A much more severe leak occurred in Eastern Europe in the former U.S.S.R. There the Chernobyl nuclear power plant released huge amounts of radiation, which started to melt into the ground. During World War II, two Japanese cities were destroyed by nuclear bombs. We may never know how many people were killed or became ill from these human-caused disasters.

Three Mile Island

Natural Disasters

Lightning strikes and flames race through brush and forest, leaping from **vegetation** to homes without stopping. Raging winds rip homes from the ground or **level** them where they stand. Weary neighbors wait on rooftops as rescuers navigate flooded neighborhoods in boats.

People frantically dig through the rubble, trying to save friends, family, and strangers after the earth shakes and thrusts. Earthquakes collapse entire cities, leaving thousands trapped or worse.

Volcanoes spew huge amounts of hot ash and soot, burying nearby towns. They spit fire and rain rocks. Hot lava flows from the top and down the sides of the volcano. The heat melts snow instantly. The water mixes with rocks, lava, and soil, sending walls of steaming mud over villages below. Moviemakers with all their special effects would be challenged to match the violent fury and visual spectacle of a major eruption.

Weather or Not?

Many natural disasters are caused by weather. The right combination of temperatures, wind, and **precipitation** can produce huge natural disasters. Thunderstorms, tornadoes, and hurricanes blow across the land, taking lives and destroying land and property. Blizzards bury towns under thick blankets of snow. Too much precipitation brings floods, which can wash away entire communities. A lack of water brings **drought** to plants, animals, and people.

Natural disasters claim lives, destroy homes, and change the face of the Earth. The weather can be disastrous!

2 Ready to Rumble?

Have you ever been startled by a sudden thunderclap? Perhaps you've jumped when a bolt of lightning struck nearby. Maybe you've even wondered if you would float away in a rush of gushing rainwater.

Most thunderstorms bring necessary rain and cause very little damage. Once in a while, however, a powerful thunderstorm can crush an area with its fierce rains, pounding hail, and sizzling flashes of light.

Rain and Hail

Do you know what causes a thunderstorm? Warm, moist air moves upward into the cold air higher up in the **atmosphere**. As water vapor rises, it forms droplets, cools, and freezes into ice particles. Huge, billowy clouds form and grow as moisture collects in them.

The water and ice particles continue to collect moisture until they become too heavy. When a cloud can hold no more moisture, a **cloudburst** occurs. The ice particles melt as they pass through warmer air. Heavy rains pound the earth.

Try This!

Have an adult boil water in a pan. Watch the water **evaporate** as it turns to steam and rises into the air. Place several ice cubes in a metal pie pan. Hold the pan over the steam. What happens?

You should notice water droplets forming on the bottom of the pan. The steam is **condensing** back into liquid water. Eventually the drops will get so big and heavy that they will fall back into the pan just like rain falling from clouds.

Ice particles that don't melt hit the ground as hailstones. Most hail is pea size or smaller, but it can be as large as a grapefruit. Some of the hailstones are continually blown back up into the clouds by strong **updrafts** of wind. The ice balls continue to pick up more moisture in the cold air, which freezes around the hailstones. If you cut a large hailstone in half, you would see several layers of ice. Extremely large hail measuring more than 17 inches across and weighing as much as 4 pounds has been discovered. Large hailstones can leave dents in cars and on rooftops. Entire fields of crops have been beaten to the ground by the raging balls of ice.

Where Does the Thunder in the Thunderstorm Come From?

When lightning heats the air quickly, the air **expands**. The particles in the air collide as they fly outward in all directions. This causes the booming sound of thunder.

Flash Floods

Severe thunderstorms can cause flash floods. These are floods that occur when a large amount of rain falls in a small area over a short period of time. This happens when a slow-moving thunderstorm takes a while to pass over an area or when a series of storms passes over the same spot.

Isolated rain showers are those that happen in a small area. These isolated showers can be very heavy, so they are also called *downpours* or *cloudbursts*.

As the rain beats down during a downpour, water collects faster than the ground can soak it up. Streets, gutters, canyons, and **ravines** become **spillways**. Water races to the lowest places in the area. These low areas quickly fill and spill over.

Flash floods are the number one weather-related cause of death. Creeks, streams, and roads become roaring rivers in minutes. Moving water can easily carry cars away. Homes and businesses can flood with little warning. People caught in the flooding water become trapped.

Sometimes it's not even raining where the flood hits. Heavy rains on higher ground can cause rushing water to flow toward lower spots nearby. These lower areas are suddenly flooded by incoming water.

In 1976, more than 145 people were killed when a flash flood hit Colorado's Big Thompson Canyon, a popular camping and hiking area. A thunderstorm dropped 14 inches of rain in just hours. Water raced down the steep canyon slopes into Big Thompson Creek. In minutes, the creek rose 19 feet, sweeping through campgrounds and motels and over roadways. People were carried away in their cars and washed away on foot. There was little warning and no help for the flood victims.

Flash flooding occurs all over the world. In 1985, Cheyenne, Wyoming, saw six inches of rain fall in just three hours. It cost this United States city $61 million to repair the damages. Twelve people died. In 1995, two rivers in South Africa spilled over following heavy rains. More than 700 homes were washed away, and more than 130 people were killed. A year later, a campsite in Spain was crushed by a wall of water and mud 12 feet high. Cars and trailers were dragged down a hill by the rushing wall.

Big Thompson flood

Lightning

Precipitation isn't the only danger in a thunderstorm. Lightning can be an equally disastrous force.

Opposites Attract

Opposite electrical charges are attracted to each other. So positive charges are pulled toward negative charges and negative charges are pulled toward positive charges.

As moisture freezes and thaws, ice particles move between the lower and upper parts of a cloud. Electrical charges are produced. The positive and negative charges separate. When these charges are attracted, or pulled, to charges in other clouds or on the ground, they jump from cloud to cloud, cloud to ground, or ground to cloud. This is the bolt of lightning you see zapping across the sky or sizzling between a cloud and the ground.

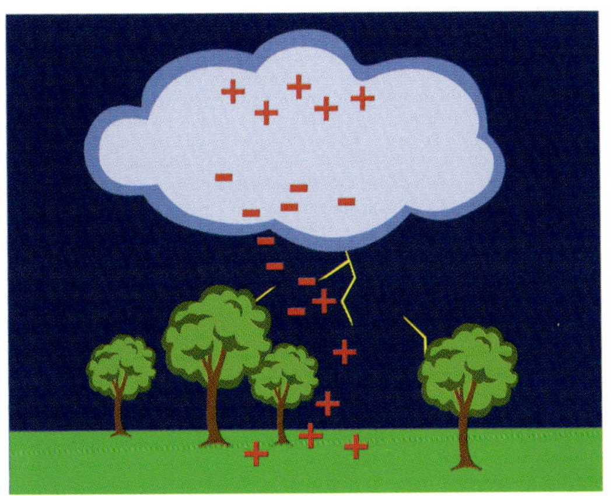

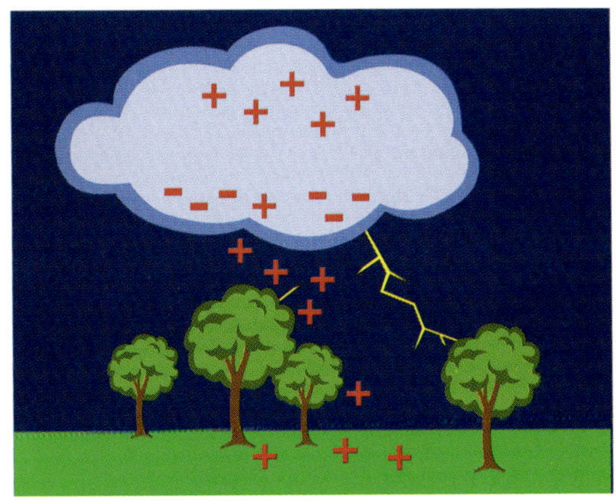

Lightning travels at a speed of over 60,000 miles per second. With temperatures of 30,000 to 50,000°F, lightning is hotter than the surface of the sun. This speed and heat make lightning very dangerous.

When lightning strikes a person, it can kill. When it hits a building, huge amounts of electricity travel through pipes and wires. People have actually been killed by lightning while talking on the telephone!

This forest fire in Oregon was caused by lightning.

Fire!

Fires caused by lightning can be devastating. Many fires in the United States are started by a single bolt of lightning. About one in three forest fires is caused by lightning.

Small fires are nature's way of thinning out vegetation and giving trees and other plants room to grow. But when these small fires rage out of control, entire forests can be destroyed.

3 Terrible Twisters

The story of Dorothy and her journey to Munchkin Land is probably the most famous tornado story. While *The Wizard of Oz* is fictional, a real tornado does have the strength to pull up a house and drop it somewhere else. But when the storm has passed, there won't be any ruby slippers, dancing scarecrows, or cowardly lions. In fact, what's left behind may be scarier than the wicked witch.

direction, the winds can twist together like a rope. These rotating, or spinning, winds form a funnel cloud. If this funnel cloud touches the ground, the tornado will whirl across the land, destroying everything in its path.

Funnel Clouds

Severe thunderstorms can create tornadoes. Warm air rises into the storm. As that air rises, more air rushes in to take its place. This causes the thunderstorm to draw in and feed off the warm, wet air.

This warm air rises and punches through colder air above. When the rising air collides with winds moving in a different

Measuring Tornadoes

The strength of a tornado is measured by its wind speed. The Fujita Scale rates tornadoes by wind speed and how much damage they cause.

Seven or eight of every ten tornadoes are considered weak. Mobile homes, storage sheds, and trees suffer the most from these F0 tornadoes.

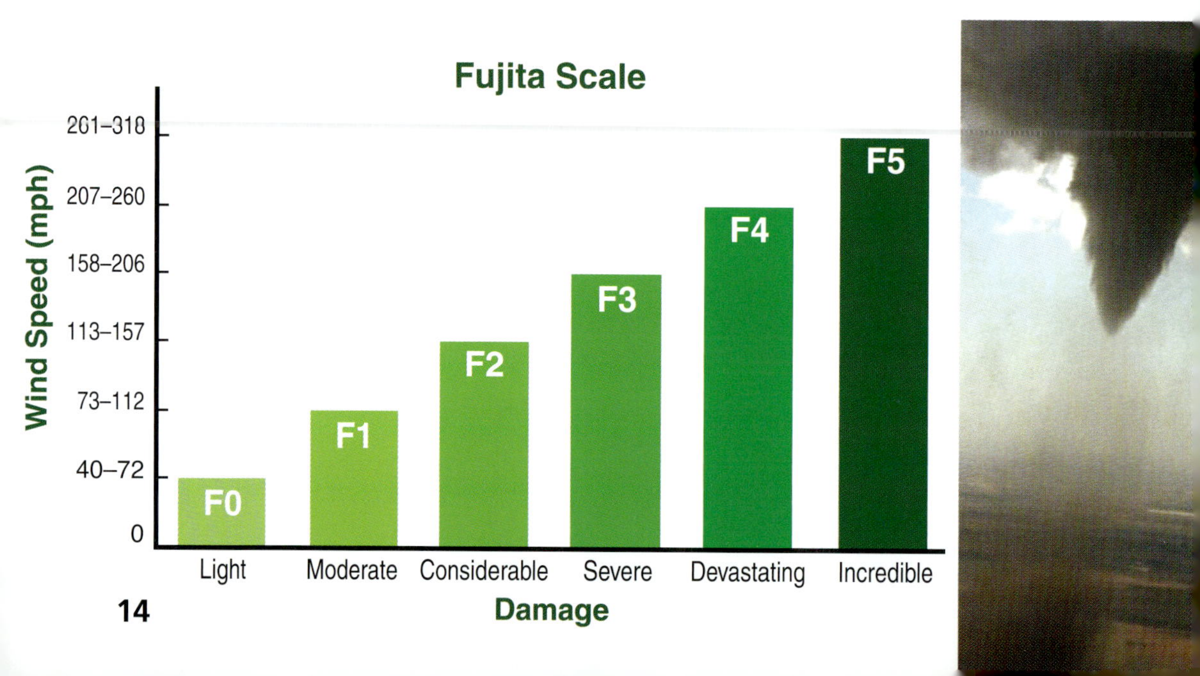

Stronger tornadoes make up one to three out of every ten. These F1, F2, or F3 storms can bring up to 200-mph winds. Homes are leveled, cars are thrown, and brick walls are blown over. Most deaths in severe tornadoes are caused by flying or falling **debris**.

With winds as high as 300 mph and a path nearly a mile wide, F5 tornadoes can cause the total destruction of an area. Entire buildings are thrown. Trains are blown off tracks. Flying debris scrapes the bark off tree stubs. Luckily, there are only a handful of extremely violent tornadoes each year.

Twister Terror

Fewer people are killed by tornadoes today than 50 years ago. This is due to the ability of **forecasters** to recognize, predict, and warn the public when tornadoes threaten. Buildings have also been made stronger to help them survive violent winds. Still, these windstorms strike terror in the hearts of many, and the stories of real-life tornadoes are frightening.

In 1931, a tornado in Minnesota picked up an 83-ton train car and the more than 100 people riding in it. The twister dropped the train car more than 80 feet away. What a ride!

Jarrel; Texas, tornado

Path of Tristate Twister

One of the worst tornadoes ever recorded happened in 1925. It was called the Tristate Twister because it hit parts of Missouri, Illinois, and Indiana. This monster tornado lasted about four hours—ten times longer than most. It demolished three towns and leveled nearly half of two others. The Tristate Twister left a path of rubble more than 200 miles long and claimed about 700 lives.

Forty-nine tornado warnings and eight twisters that actually touched down made May 27, 1997, a day that people in Jarrel, Texas, will always remember. Rated as an F5, one tornado's winds were estimated at more than 260 mph. In less than ten minutes, 27 people were killed.

The average tornado lasts less than 20 minutes. It travels less than five miles with a path no wider than a football field. But the damage it can do is definitely disastrous!

4

In the Eye of the Storm

Have you ever seen a boat slam into an apartment building? Or watched a huge wave of water bury an entire block of houses? When hurricanes hit land, they send a wall of water called a *storm surge* as high as 30 feet above the normal water line. The surge can stretch for 50 miles along the coastline, wiping out everything in its path.

Forecasting Tornadoes Versus Hurricanes

Tornadoes may have nature's strongest winds, but hurricanes last longer and cover a much larger area. The life of a tornado can be measured in minutes. The life of a hurricane is measured in days or weeks.

One of the scariest things about a tornado is that you never know exactly where, when, or if it's going to happen. Forecasters can tell ahead of time if a tornado *might* happen. They can predict where it *might* happen. They can even determine the approximate time of day when it *might* happen. But until it *does* happen, they never know for sure. When a tornado actually forms, the people in its path might have only minutes to take shelter.

Hurricanes, however, give forecasters much more warning. A hurricane starts as a thunderstorm near the equator. As warm ocean waters feed the storm, the winds begin to swirl at high speeds. With winds up to 38 mph, this storm is called a *tropical depression*.

As the storm keeps gathering strength, the winds grow stronger. When they reach 39 mph, it becomes a tropical storm. The storm officially becomes a hurricane when the winds reach a speed of 74 mph or more.

Forecasters can watch storms turn into hurricanes for a week or more. Watching the storm grow, forecasters predict where the hurricane will hit several days ahead of its arrival. When the storm is 8 to 16 hours away, warnings are put out and **evacuations** are advised or ordered.

A Storm with Many Names

Hurricanes have different names in different parts of the world. When they form near the Atlantic Ocean, the storms are called *hurricanes*. Over the Indian Ocean, they're called *cyclones*, while in the Pacific Ocean, they're known as *typhoons*. In Australia, many people call them *willie willies*.

A Hurricane!

As warm air rises and cold air sinks, the weight of the atmosphere changes. This causes low **air pressure**. Just like water going down a drain, the violent storm swirls around the area of low pressure. The winds swirl counterclockwise north of the equator and clockwise south of the equator.

The most violent part of the storm is on the edges, or wall, of the hurricane. As the outer wall of the hurricane approaches an area, wind, waves, and rain smash into communities. Land and property are destroyed. People and animals are injured or killed.

Inside the tight, twisting wall of wind, the center, or eye, of the hurricane is strangely calm. As it passes over an area, the mild weather may make it seem as though the storm is over. But the winds pick up again as the second half of the storm arrives.

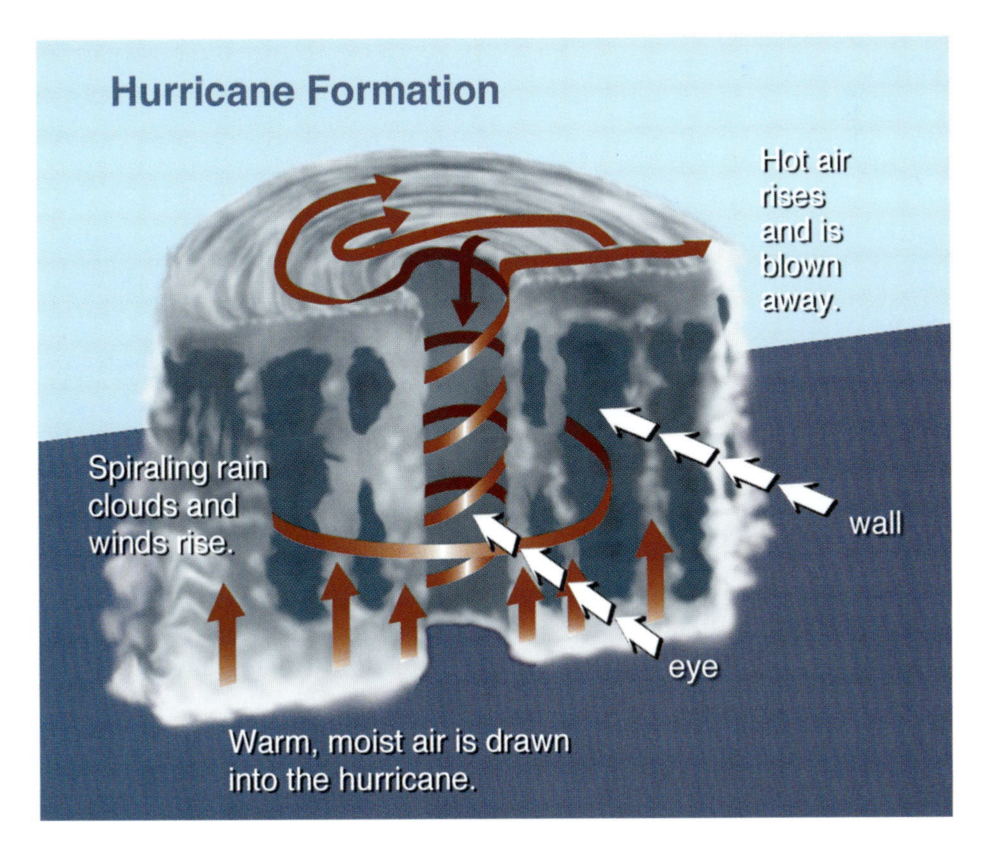

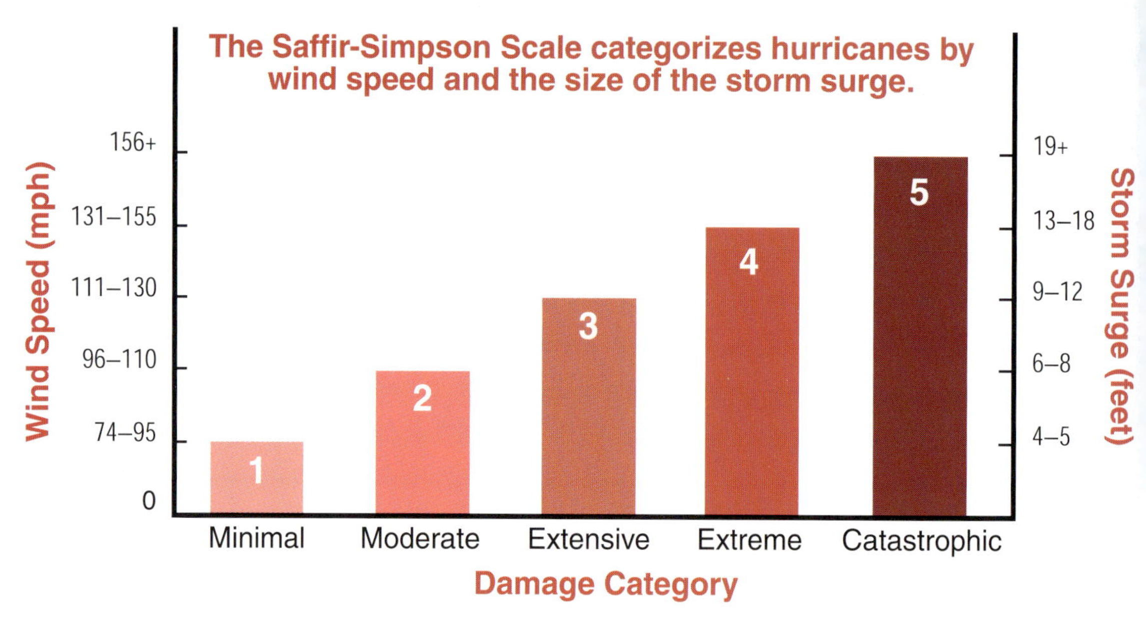

Naming Hurricanes

All tropical depressions, storms, and hurricanes are studied by **satellite**. Hurricanes are given names to help identify storms and track them as they move. In the past, there have been different systems for naming these storms. Some were based on saints' names or the language of the countries where the hurricanes hit. Others were given female names only. Today, six lists of names are rotated from year to year. The names are in alphabetical order, but the letters Q, U, X, Y, and Z aren't used.

Damage left behind after Hurricane Camille

Retired Hurricane Names

If a storm is extremely destructive or costly, the name is **retired**. A new name is then chosen and added to the lists. Some of the retired hurricane names are Alicia, Bob, Carmen, Donna, Mitch, and Roxanne.

Hurricane Horror Stories

Forecasting storms and putting out warnings save lives. But this doesn't lessen the power of a hurricane. Residents of southern Florida learned that in 1992 when Hurricane Andrew came calling. One of the most damaging hurricanes ever, Andrew tore into Miami with 165 mph wind gusts 35 miles across. Two hundred and fifty thousand people lost their homes there. But Andrew wasn't finished. Continuing across the coast, Andrew's winds dropped to 140 mph when the hurricane hit Louisiana two days later. Hurricane Andrew ended the next day. Thirty thousand people were left without homes. Half the state's important sugar crop was destroyed.

In 1988, Hurricane Gilbert struck Jamaica, Mexico, and the United States. This Category 5 storm caused billions of dollars of damage to communities in Mexico and killed more than 300 people.

Hurricane Camille began in August of 1969. With winds as high as 200 mph, the storm hit the U.S. Gulf Coast. Flooding in Mississippi, Louisiana, and even Virginia caused more than 200 deaths. Camille was one of the costliest disasters in U.S. history.

Hurricanes Andrew, Gilbert, and Camille caused extreme damage to property and loss of human lives. Because of this, these three are among those names on the retired list.

5

Water, Water Everywhere!

Water is essential to all life. Humans, plants, and animals depend on it. But unless you're a fish, *too much* water can be a big problem!

Floodwaters rushing through city streets damage cars and buildings. Unexpected walls of water sweep away homes. People are injured by debris in the water. Some are trapped by the rising flood.

Flooded treatment plants and broken pipes can spread **sewage**, making entire populations sick. There may be water everywhere but not a drop to drink if it's polluted.

Racing across farmland, floods can strip fields of rich **topsoil**. Crops drown, and important **nutrients** are washed away. **Erosion** caused by floods can change the landscape of entire communities.

River Floods

Too much rain or melting snow can cause rivers to fill and overflow. These floods are usually more gradual than the flash floods that come with heavy storms. It may take weeks or months of rain or snow to cause flooding. But when the ground around the river cannot **absorb** any more moisture and the river spills over its banks, the flooding begins.

In 1993, it rained for months in the Midwest. The Mississippi River began to rise. The ground was soaked. As the rain continued, the water rose until finally it spilled over, causing flooding in nine states. Fields became lakes. Roads were covered with raging waters. Cities went without clean water for days and weeks. Thousands of homes were destroyed. Towns were demolished. Eventually the rain stopped, but the flood had left its mark on the Midwest.

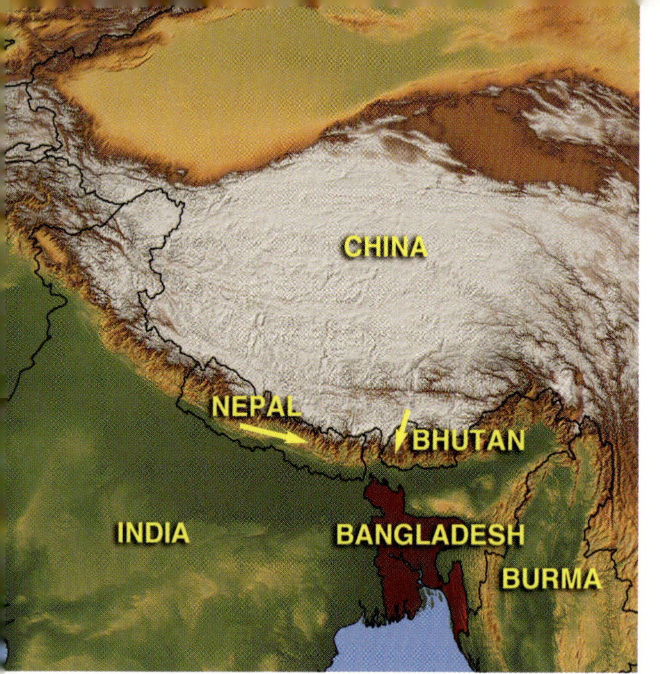

Coastal Floods

Hurricanes and tropical storms can cause flooding along coastal land. Entire beaches and coastal towns are swept away by great waves of water. Most deaths from hurricanes occur because of flooding.

Bangladesh, a small country in Asia, has been struck by several devastating coastal floods. In 1970, 300,000 people were killed during a hurricane. The country was hit again in 1991 and 1993. Over a million people died in a storm surge in 1993 when a 15-foot wall of water washed over the land.

Urban Floods

Urban, or city, flooding happens when heavy rain falls in an area with many buildings and paved roads and parking lots. The water can't be absorbed by the little remaining soil, so it floods the city. Water spills into streets and basements. Entire buildings and homes are ruined.

In 1999, heavy rains in several cities in Africa caused urban flooding. Thousands of residents were forced to evacuate their homes. At least eight people were killed. As Africa continues to develop more towns and cities, increased urban flooding is expected in the future.

Urban flooding in Reno, Nevada

Flooding after a monsoon in India

Monsoons

In Southern Asia, monsoon rains arrive every summer. During this time, seasonal winds called *monsoons* blow across the Indian Ocean, picking up moisture. The monsoons bring heavy downpours. These storms are important because they bring needed rain for crops. When the monsoons are too heavy, however, rivers overflow and wash away the crops. Villages are flooded and destroyed. Lives are lost.

In the summer of 2002, the rains were especially heavy. Monsoon rains drove rivers over their banks. The floodwaters raged, running down the **foothills** of the Himalayas and spreading from Nepal through India and into Bangladesh. Five hundred and eighty people were killed. Half of Bangladesh was underwater. As floodwaters **receded**, nearly 200 million people were left homeless.

But as communities repair themselves, they once again await next year's rains. For the people of Asia who depend on water from monsoons, these natural disasters are welcomed more than feared.

Seasonal Storms

The word *monsoon* comes from the Arabic word *mawsin,* which means "season." The storms earned this name because they are expected every year during the same season.

6

Dry as a Desert

Does your family have a vegetable garden? Is your yard covered with grass and blooming with flowers in the spring and summer? When it doesn't rain for a while, what happens? Do you turn on the sprinklers or water the garden with a hose? What happens if you forget?

Chances are that without water, the grass will dry up and turn brown. The plants will droop and wilt. Eventually everything living will shrivel up and die. Your beautiful yard will become a dry, crumbly desert.

Imagine that happening to *all* the crops and **grazing** land across an entire region. This is called a *drought*. Severe drought can be a long and painful disaster. **Wells**, streams, and lakes dry up. Animals die of thirst or starve when their food shrivels up. People who live in areas that depend on crops or livestock as a main food source may also die. A **famine** may result.

Drought doesn't just affect farmers. People in cites feel the effects too. Orders to restrict water use can cause problems. Food prices may go up since surviving crops will be **scarce**. With so little moisture, small fires can become deadly in minutes. Homes and businesses can quickly go up in flames.

What Causes a Drought?

Something that isn't very heavy is sometimes said to be "as light as air." But air is actually pretty heavy. The weight of the air pushing down on the Earth changes depending on temperatures and winds. An area of high pressure exists when the air becomes heavier. A low-pressure area results from lighter air. High air pressure in an area can prevent storms from passing over the region, so precipitation becomes scarce.

Another factor in drought is the temperature and movement of ocean water. Oceans near the equator pump moisture into the air. This moisture is carried by the wind and causes storms all around the world. Since warm air holds more moisture than cold air, changes in the flow of warm ocean moisture can cause a drought.

The Dust Bowl

During the 1930s, parts of the United States experienced extreme drought. For eight years, Texas, Oklahoma, Kansas, and Colorado received little or no rain. No crops grew. The soil dried up. Winds began to blow. The dry, dusty topsoil swirled around in the air, dropping everywhere. Homes, businesses, and people were soon covered with this "dust." Damage from dust storms spread into other states, such as Nebraska, North and South Dakota, Wyoming, and Montana.

Many families had to leave their homes and find new lives elsewhere.

The Sahel Shore

Sahel is the Arabic word for "shore." The Sahel region forms the edge, or shore, of the Sahara Desert. The Sahel covers land in parts of Senegal, Mauritania, Mali, Burkina Faso, Niger, Nigeria, and Chad.

Disastrous Drought

Parts of Africa have been hit hard by drought. Entire populations have lost their way of life due to drought in the Sahel region and Ethiopia. This region is normally very dry. But between 1968 and 1974, the little rain the area usually receives dropped off to almost nothing. Crops were destroyed. Most of the livestock died. The rest ate the last of the wild grasses, leaving no plants to seed new growth when the rains returned. Dry conditions and no plant life helped the wind strip away **fertile** topsoil, creating desert land. The Sahara Desert grew by over 60 miles. More than 100,000 people died in this drought.

7 Whiteout!

Strong winds gust in your face. Heavy, blowing snow swirls around you. Freezing temperatures chill you to the bone. A blur of white surrounds you. You're caught in a blizzard!

As heavier, cold air slips under warm, wet air, it causes the moisture in the warm air to fall. If the air beneath is cold enough, it falls as snow. A calm, gentle snowfall can be quite beautiful. But mix in strong winds, colder temperatures, and more moisture, and it becomes a blizzard.

Blizzard Conditions

A snowstorm becomes a blizzard when
- winds blow at 35 mph or more
- temperatures are cold but above 10°F
- **visibility** is under 1500 feet

A blizzard is considered severe when
- winds blow over 35 mph
- temperatures are below 10°F
- there's near-zero visibility

Beware! It's a Blizzard!

While you may think of snow as a vacation day from school, it can mean disaster for many. As the snow falls, it is blown around by strong winds. The snow builds up in big piles, or drifts, along fences, buildings, parked cars, and other objects. Heavy snow can drift across roadways, burying sidewalks and streets. Entire cities may become trapped, or snowbound. People can't go to work, children can't reach school, and injured or sick people may not be able to reach medical help in time.

Cars that do make it onto roads face danger from slippery ice and **whiteout** conditions. People stuck in cars stalled by the snow risk freezing from the cold

temperatures. Walking in a blizzard can be just as dangerous. Icy winds and blinding snow can trap people in the storm.

Lost in a Blizzard

In 1931, twenty children and a bus driver were lost in a blizzard in Colorado. While trying to deliver the children to safety, the bus driver got stuck in the storm. Wind and snow blew into the bus as the children waited to be rescued. By the next morning, the winds were blowing about 70 mph, and it was -20°F. The bus was finally found that afternoon, but five of the children and the bus driver didn't survive the deadly storm.

Avalanche!

As snow drifts and **accumulates**, areas with steep slopes and **overhangs** can quickly become avalanche areas. Strong winds can blow snow over **ridge** tops, causing it to collect and form unstable overhangs. Tons of snow with no support underneath and changing temperatures cause the snow mass to become more and more unstable.

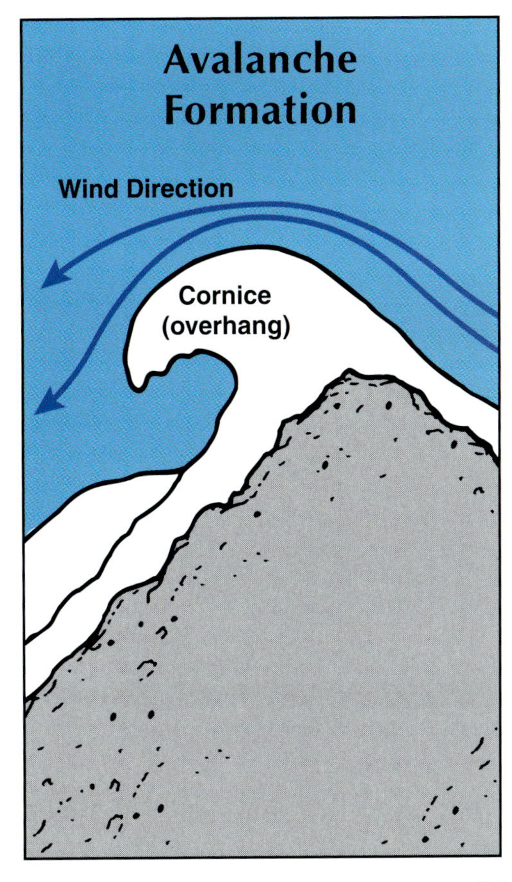

Avalanche in Alaska

Weight, wind, temperature, or an object such as a skier can cause the snowy overhang to break loose. This usually happens a day or more after the storm has passed. As the mass of snow slides down a mountainside, it gathers more snow, rocks, and other debris. Anything below is in danger of being swept away, broken, twisted, or buried.

Being caught in an avalanche can be deadly. In the United States, more people die in avalanches than in hurricanes or earthquakes. In Canada, 91 people were killed in avalanches between 1979 and 1994. Most of the victims were enjoying fun activities in the snow, such as skiing, snowboarding, or snowmobiling. Hundreds of people participating in winter activities in European countries have been killed by avalanches in past years.

Explosive Prevention

In areas where avalanches are common, experts keep a close watch and take steps to prevent the dangerous snow from causing harm. Explosives are used to bring down the hanging snow before allowing people back into the danger zone.

8

Forecast for Disaster

While scientists know much about the natural disasters that affect our world, they are still searching for better ways to predict and measure them. Today, forecasters have many tools that provide accurate information about the weather. Planes fly into hurricanes. Scientists chase tornadoes. Forecasters follow storm paths and provide up-to-date reports.

Paying attention to forecasts and taking all necessary precautions will decrease damages and lives lost. Natural disasters cannot be avoided, but we can do all that's possible to lessen their effects on our lives.

Forecasting Tools

Tools	Job
anemometer	measures wind speed
barometer	measures air pressure
computer imaging	computer program that tracks storms and shows their movement
radar	device that picks up the location of storms
rain gauge	measures precipitation
satellite	object that orbits in space and collects weather information
thermometer	measures temperature

Internet Connections and Related Reading for Natural Disasters

http://www.fema.gov/kids/dizarea.htm
This is the Federal Emergency Management Agency's disaster site for kids. Explore natural disasters through facts, activities, and stories. Learn how to protect yourself in a natural emergency.

http://wxdude.com
Let the "Weather Dude" show you the ABCs of weather. This extensive site for kids features an alphabet tour of weather, weather basics, and an online book. Information on severe storms, snow, lightning, hurricanes, and other natural dangers can be found here.

http://tqjunior.thinkquest.org/5818/index.htm
Experience "Weather Gone Wild." This site covers tornadoes, thunderstorms, snowstorms, hurricanes, and floods.

http://www.miamisci.org/hurricane
Go inside a hurricane to understand its causes and effects. Make your own weather instruments, and discover how hurricanes are tracked.

http://mos.org/sln/toe/toe.html
Visit the "Theater of Electricity" online. Read about Ben Franklin's kite experiment, learn a few lightning facts, and take a lightning safety quiz.

http://weathereye.kgan.com/cadet/flood/about.html
Keep your eye on floods at this site. Check out information on types of floods, how to prepare for a flood, and examples of these devastating disasters. Take a quiz to see how much information has flooded your brain!

Blizzard! Snowstorm Fury by Mary C. Turck. Destructive winds and blowing snow have devastated rural areas and cities throughout history. Learn about some of the worst blizzards and what to do to survive in one. Perfection Learning Corporation, 2000. [RL 4.3 IL 4–9] (5785301 PB 5785302 CC)

Flash, Clash, Rumble, and Roll by Franklyn M. Branley. Young readers will discover not only what is happening during a thunderstorm but also how to stay safe during one. HarperCollins, 1985. [RL 3 IL K–3] (8453101 PB 8453102 CC)

Floods! Rising, Raging Waters by Jane Duden. What causes floods? Where do they occur? When are they disasters? Read to discover answers to these and other questions. Includes firsthand accounts of some of the most devastating natural disasters in recent times. Perfection Learning Corporation, 1999. [RL 4.2 IL 4–9] (5760601 PB 5760602 CC)

Horror from the Sky: The 1924 Lorain, Ohio, Tornado by Bonnie Highsmith Taylor. Calvin fears for his brother's life after a tornado rips through their Ohio town. Perfection Learning Corporation, 2002. [RL 2.2 IL 4–6] (3864001 PB 3864002 CC)

Hurricane! Nature's Most Destructive Force by Margo Sorenson. Cities have been destroyed and lives lost from these gigantic storms. Find out how hurricanes form and what's being done to track them. Full-color photos, technical and historical data. Perfection Learning Corporation, 1997. [RL 4.3 IL 4–9] (4946601 PB 4946602 CC)

Lightning by Seymour Simon. An introduction to lightning, one of nature's most fierce and mysterious forces. William Morrow, 1997. [RL 4 IL 2–5] (3277101 PB 3277102 CC)

The Magic School Bus Kicks Up a Storm by Joanna Cole. Ms. Frizzle's class learns about the weather when the bus becomes a flying weather station and takes the class right into the storm clouds. Scholastic, 2000. [RL 3.2 IL K–4] (3752401 PB)

Night of the Twisters by Ivy Ruckman. A fictional account of the night freakish and devastating tornadoes hit Grand Island, Nebraska, as experienced by a 12-year-old boy, his family, and friends. HarperCollins, 1986. [RL 4 IL 3–6] (8596101 PB 8596102 CC)

Snow Is Falling by Franklyn M. Branley. Describes the characteristics of snow, its usefulness to plants and animals, and the hazards it can cause. HarperCollins, 1986. [RL 2.3 IL K–3] (8607301 PB 8607302 CC)

Tornado! The Strongest Winds on Earth by Mike Graf. Learn how, when, and where these killer storms form and what we can do to protect ourselves and our property. Perfection Learning Corporation, 1999. [RL 3.9 IL 4–9] (5623801 PB 5623802 CC)

- RL = Reading Level
- IL = Interest Level

Perfection Learning's catalog numbers are included for your ordering convenience. PB indicates paperback. CC indicates Cover Craft. HB indicates hardback.

Glossary

absorb (uhb SORB) take in

accumulates (uh KYOUM you layts) increases or piles up

air pressure (air PRESH er) force or push of the air

atmosphere (AT muhs fear) air surrounding the Earth

cloudburst (KLOWD berst) sudden heavy rainfall

condensing (kuhn DENS ing) changing from a gas to a liquid

debris (duh BREE) remains of something broken down or destroyed

drought (drowt) extreme lack of water

erosion (ee ROH zhuhn) wearing away of soil by wind and water

evacuation (ee vak you AY shuhn) to remove from a dangerous area

evaporate (ee VAP or ayt) to change from a liquid to a gas

expands (ik SPANDZ) spreads out or grows larger

famine (FAM in) extreme lack of food among many people

fertile (FER tuhl) good for growing

foothill (FOOT hil) hill at the bottom of a larger hill or mountain

forecaster (FOR kas ter) person who studies and predicts weather conditions

grazing (GRAYZ ing) relating to land with grasses and plants that animals feed on

habitat (HAB i tat) place where a plant or animal lives and grows

level (LEV uhl) to knock down

nutrient (NOO tree ent) something a living thing needs to grow

overhang (OH ver hayng) object that sticks, or hangs, over the edge of something

precipitation (pree sip uh TAY shuhn) moisture that falls from the clouds, such as rain and snow

radiation (ray dee AY shuhn) energy waves that are harmful when exposed to humans

ravine (ruh VEEN) small, narrow valley often formed by running water

receded (ree SEE duhd) backed away; fell back

retired (ree TEYERD) no longer used

ridge (rij) range, or group, of hills or mountains

satellite (SAT uh leyet) object or vehicle that orbits in space and can record information from its position

scarce (skairs) not plentiful; not in large quantity

sewage (SOO ij) waste material

spillway (SPIL way) passageway for excess water to run over or around objects such as buildings or mountains

topsoil (TOP soyl) soil at the surface where most plant roots grow

updraft (UHP draft) upward movement of air

vegetation (vej uh TAY shuhn) plant life in an area

visibility (viz uh BIL uh tee) ability to see

well (wel) pit or hole dug underground to store water

whiteout (WEYET owt) very little visibility due to a blizzard (see separate entry for *visibility*)

Index